Forecasting Science and Technology for the Department of Defense

John W. Lyons, Richard Chait, and James J. Valdes

Center for Technology and National Security Policy
National Defense University

December 2009

The views expressed in this article are those of the authors and do not reflect the official policy or position of the National Defense University, the Department of Defense, or the U.S. Government. All information and sources for this paper were drawn from unclassified materials.

John W. Lyons is a Distinguished Research Fellow at the Center for Technology and National Security Policy (CTNSP), National Defense University. He was previously director of the Army Research Laboratory and director of the National Institute of Standards and Technology. Dr. Lyons received his PhD from Washington University. He holds a BA from Harvard.

Richard Chait is a Distinguished Research Fellow at CTNSP. He was previously Chief Scientist, Army Material Command, and Director, Army Research and Laboratory Management. Dr. Chait received his PhD in Solid State Science from Syracuse University and a BS degree from Rensselaer Polytechnic Institute.

James J. Valdes is a Senior Research Fellow at the National Defense University's Center for Technology and National Security Policy and the Army's Scientific Advisor for Biotechnology. Dr. Valdes received a PhD in neuroscience from Texas Christian University and was a postdoctoral fellow at the Johns Hopkins Medical Institutes. He has published more than 120 papers in scientific journals and was a 2009 Presidential Rank Award winner.

Acknowledgments

The authors gratefully acknowledge Dr. Thomas Killion, Army S&T Executive, for his continuous support and interest in this area, and Ms. Cheryl Loeb and Ms. Ewelina Tunia for their astute assistance in preparation of the manuscript.

Defense & Technology Papers are published by the National Defense University Center for Technology and National Security Policy, Fort Lesley J. McNair, Washington, DC. CTNSP publications are available at http://www.ndu.edu/ctnsp/publications.html.

Contents

Introduction

Since World War II, predictions of science and technology for military applications have occurred periodically. A study chartered by the Army Air Force[1] predicted in 1947 a broad range of developments in aeronautics and air power and has been a model for such forecasts ever since. Projections in science and technology have been issued for many years by the National Research Council (NRC) of the National Academies, which publishes decadal studies for specific disciplines. Such studies for astronomy and astrophysics, for example, go back to at least 1964.

An important task of DOD science and technology (S&T) programs is to avoid technological surprise resulting from the exponential increase in the pace of discovery and change in S&T worldwide. The nature of the military threat is also changing, with the result being new military requirements, some of which can be met by technology. Shaping the S&T portfolio requires predicting and matching these two factors well into the future. Some examples of technologies that have radically affected the battlefield include the Global Positioning System coupled with inexpensive, handheld receivers; the microprocessor revolution, which has placed the power of the Internet and satellite communications in the hands of soldiers in the field; new sensing capabilities such as night vision; and composite materials for armor and armaments. Some of these technologies came from military S&T, some from commercial developments, and still others from a synthesis of the two sectors, but all were based on advances in the underlying sciences. Clearly, leaders and planners in military S&T must keep abreast of such developments and look ahead as best they can.

In the Department of Defense (DOD), the last series of forecast studies was done in the 1990s.[2] In 2008, National Defense University's Center for Technology and National Security Policy (CTNSP) assessed the Army's STAR 21 (Strategic Technologies for the Army of the Twenty-First Century) study,[3] in which the basic and applied sciences were assessed and forecast as separate and discrete disciplines. Future capabilities were discussed in a separate set of STAR 21 volumes on systems. In general, the technologies of individual systems were not discussed with reference to the underlying sciences. This

---

[1] Theodore von Karman, *Toward New Horizons* (Washington, D.C.: United States Army Air Force, 1945). The study, performed by the new Army Air Force Science Advisory Group, chaired by Von Karman, charted the way ahead for air power for the United States. The history of this study is in: Michael H. Gorn, editor, *Prophecy Fulfilled, 'Toward New Horizons' and its Legacy* (Washington, DC: Air Force Historical Studies Office, 1994), available at <http://www.airforcehistory hq.af.mil/Publications/authorindex htm.
[2] Board on Army Science and Technology, Commission on Engineering and Technical Systems, National Research Council, *STAR 21—Strategic Technologies for the Army of the Twenty-First Century* (Washington, DC: National Academy Press, 1992); Naval Studies Board, National Research Council, *Technology for the United States Navy and Marine Corps, 2000-2035* (Washington, DC: National Academy Press, 1997); Air Force Scientific Advisory Board, *New World Vistas, Air and Space Power for the 21st Century* (Washington, DC: Department of the Air Force, 1995).
[3] John Lyons, Richard Chait, and Jordan Willcox, *An Assessment of the Science and Technology Predictions in the Army's STAR21 Report,* Defense & Technology Paper 50 (Washington, DC: Center for Technology and National Security Policy, July 2008).

separation of future capabilities from the underlying S&T forecasts was true for the studies of all three services.

Forecasting the future 25 or 30 years out is a chancy business at best, and one should not expect perfection. The CTNSP assessment of the Army's STAR 21 predictions revealed that, after 15 years, about a quarter of the predictions were right on target as of 2008, some were overly optimistic, some were too conservative, some were wrong, and some important developments were missed completely, as illustrated by the following examples:[4]

Correctly predicted:
- The information explosion
- Speed of computer operations
- Biosensors
- Organic and resin matrix composites
- Diesel turbo compounding

Progress overestimated:
- Information saturation
- Automated target recognition
- Metal matrix composites
- Electromagnetic gun technology

Progress underestimated:
- Distributed processing
- Computer memories
- Functional materials
- Manufacturing at nanoscale

Some wrong predictions:
- Development of a battle control language
- Free electron lasers for destroying missiles
- Tungsten to replace depleted uranium in armor and penetrators

Significantly, the STAR 21 study did not anticipate the impact of the Internet and the World Wide Web, the advent of international fiber optic links, the proliferation of personal computational devices, nor the spread of wireless technology. Revolutions are difficult to predict; evolutionary changes are easier. The imperfections of the STAR 21 study did not, however, lessen its usefulness. It served to educate the senior leadership in the Army about the exciting nature of S&T. A copy of the summary report was sent to all

---

[4] Selected from John Lyons, Richard Chait, and Jordan Willcox, *An Assessment of the Science and Technology Predictions in the Army's STAR21 Report,* Defense & Technology Paper 50 (Washington, DC: Center for Technology and National Security Policy, July 2008).

general officers.[5] The study gave additional credibility to budget proposals for S&T and broadened the horizons and challenged the thinking about the future of those Army scientists and engineers who sat in on the deliberations of the NRC study committee.

RISK FACTORS

There are many reasons why a forecast may go astray. One is the quality of the expertise on the study committees. Included in this may be a too-strong bias or even conflict of interest on the committees. For example, some members may try to sway the group to emphasize their particular areas It is important to appoint people with broad experience and perspective, and it would be helpful to include some futurists to stimulate imaginative thinking. Another reason may be attempting to cover too many areas in too short a time. In STAR 21, long-term basic research was separated from shorter-term technology forecasts. The CTNSP retrospective analysis suggests that the two should be studied together in order to reduce redundancies and to improve the understanding of how the basic and applied studies reinforce each other. Finally, some things, especially those with truly disruptive potential, often just cannot be anticipated, either in S&T itself or in the changing environment.

The CTNSP report ended with recommendations to update the Army study, combine forecasts of the sciences with the technologies, and conduct at least a portion of such studies jointly with the Navy and the Air Force.

This paper begins with a discussion of recent trends in S&T, particularly how various disciplines have converged to produce new capabilities. There follows consideration of how a new series of studies might be conducted with an eye to taking into account such convergences. We offer a detailed set of recommendations on the organization and management of a series of tri-service studies.

---

[5] Private communication from LTG Mal O'Neill (US Army ret.), December 3, 2007.

Background

In many cases, forecasting is an implicit part of near-term program planning, defined as the direction a program will take over 5 years (the Federal budget is supposed to cover the next budget year plus 4 additional years). In planning the work of a laboratory, one has to take into account the mission of the organization and the expressed and anticipated needs of the users of the results. Planners also have to assess the laboratory's strengths and ability to mount new programs in a timely way, while taking into account a vision of what the trends in S&T portend. These trends have to be assessed and then predictions made for the future,[6] but formal, long-range S&T forecasting has not often been attempted within DOD. On occasion, technology leaders may sponsor detailed studies of advances in S&T many years into the future. The three studies, STAR 21 and the Navy and Air Force studies, in the 1990s were done by outside experts. Two were done on contract to the National Academies' NRC; the third was done by an advisory board consisting of outside experts but convened by the service secretary and chief of staff.

In past studies there has been a tendency to make projections in the sciences in "stovepipes," restricting the subject matter analysis to one scientific discipline. Thus, the NRC has been conducting decadal studies in such areas as astronomy, atmospheric sciences, and chemical engineering. Because the trend in research is toward multidisciplinary and transdisciplinary subjects, forecasts will have to follow suit.

---

[6] John W. Lyons, *Assessing and Predicting for Army Science and Technology,* Defense & Technology Paper12, (Washington, DC: Center for Technology and National Security Policy, March 2005).

Convergences

Assessments may begin by looking back in time to chart the development of the disciplines of interest leading up to an evaluation of the current state of the art. One example of assessing the evolution of science and technology is our series of papers on the development of four successful Army systems as the products of a study called Project Hindsight Revisited.[7] In each of the four systems, new technologies came together to enable new capabilities. The improved armor of the Abrams main battle tank arose from developments in processing depleted uranium, in design of the tank silhouette, in new or improved materials such as composites, and new welding technology for fastening the plates that surround the hull and the turret. Similar developments occurred for the tank's 120mm gun, the 120mm kinetic energy round (M829A3), and gun accuracy technologies The same picture emerged from studies of the Apache attack helicopter and the Stinger and Javelin man-portable missile systems.

The advent of radar can be traced back to the convergence of the science of electromagnetic radiation and its behavior in interacting with materials, the development of microwave generators, antennae, transmitters, power supplies, and displays.[8] The various technologies converged only in 1936.

More recently, advances in solid-state physics, processes for making microelectronics, fiber optics, and computer technology made possible the technologies that have produced worldwide, broadband digital networks. Progress in telephony led to handheld wireless devices that are both telephones and computers, and access to the Internet is now available through telephone networks, bringing this particular technology convergence full circle.[9]

Two examples of convergence in the life sciences are especially illustrative. In the early 1980s, scientists began to think about designing sensors that did not require a priori

---

[7] Richard Chait, John Lyons, and Duncan Long, *Critical Technology Events in the Development of the Abrams Tank,* Defense & Technology Paper 22 (Washington, DC: Center for Technology and National Security Policy, 2005); Richard Chait, John Lyons, and Duncan Long, *Critical Technology Events in the Development of the Apache Helicopter,* Defense & Technology Paper 26 (Washington, DC: Center for Technology and National Security Policy, 2006); John Lyons, Richard Chait, and Duncan Long, *Critical Technology Events in the Development of the Stinger and Javelin Missile Systems,* Defense & Technology Paper 33 (Washington, DC: Center for Technology and National Security Policy, 2006); John Lyons, Richard Chait, and Duncan Long, *Critical Technology Events in the Development of Selected Army Weapons Systems,* Defense & Technology Paper 35 (Washington, DC: Center for Technology and National Security Policy, 2006); Richard Chait, John Lyons, Duncan Long, and A. Sciarretta, *Enhancing Army S&T—Lessons from Project Hindsight Revisited,* (Washington, DC: Center for Technology and National Security Policy, 2007).

[8] Timothy Coffey, Jill Dahlburg, and Elihu Zimet, *The S&T Innovation Conundrum,* Defense & Technology Paper 17, (Washington, D.C.: Center for Technology and National Security Policy, August 2005).

[9] The impact of these innovations on, for example, the world of business is described in Thomas L. Friedman, *The World is Flat, A brief History of the Twenty-First Century, Release 2.0,* (New York: Farmer, Strauss and Giroux, 2006).

knowledge of the substance to be detected and could in fact be trained to mimic physiology in an artificial system. Since the most sensitive natural detectors were the olfactory systems of animal species, such as pigs and dogs, it was clear that an understanding of the molecular physiology of ligand-receptor interactions, as well as the integrative properties of neuronal systems, would offer a solution to this challenge. An early concept for the design of such sensors involved coupling physiological receptors with physical optical and electronic sensors to create what is now called a "biosensor."

The technological bottlenecks to achieving a true biosensor were many. Receptor biochemistry was in its infancy; a number of receptors had been purified, but most could not function outside their native membranes and were very labile. A few receptors had been cloned and produced in small quantities, but biomanufacturing technology was essentially nonexistent. Studies therefore focused on isolating receptors from animal tissues and coating crude homogenates onto microsensor platforms. A second major barrier was a lack of understanding at the molecular level of the interactions between organic receptors and their nonbiological, microsensor supports. Third, there was very limited ability for the de novo design of novel receptors that could be tailored to have specific functional properties, nor was there much success with artificial receptors, such as molecularly imprinted polymers or bioengineered antibodies. Finally, the materials, micromachining techniques, and computational software required to design and assemble an integrated biosensor were nonexistent or, at best, inadequate.

Over the course of two decades, a series of technological convergences occurred that radically reshaped the nascent field of biosensing in a nonlinear manner. These technological convergences represented a transdisciplinary approach to science, one in which basic concepts in one discipline would fundamentally alter approaches to a seemingly unrelated discipline. First, molecular biological techniques allowed the cloning of receptor proteins and their production in large quantities, which in turn permitted detailed studies of their structure and function. Driven by developments in prosthetics and tissue engineering of artificial organs, the interface between receptors and their artificial supports became the object of intense study. Further, detailed studies of the receptors coupled with new computational methods resulted in the design of protein motifs with specific functional characteristics, and ultimately to de novo design of artificial receptors. Finally, nanotechnology and its offshoots provided unprecedented control over material properties using both directed and self-assembly techniques.

The coincident maturation of many seemingly unrelated disciplines such as protein chemistry, molecular biology, nanomaterials, metabolic engineering, biomanufacturing, systems engineering, informatics, micromachining, and neurobiology was equal parts serendipity and design. The result of that maturation was an accelerated leap forward of biosensor technology and opportunities in such areas as industrial process control, biological computing, and personal medicine that could not have been imagined in the absence of technological convergence.

Another, more famous, example of scientific and technological convergence was the confluence of organic chemistry, physics, genomics, and information technology that has transformed our understanding of living systems. In 1953, James Watson and Francis Crick announced that they had elucidated the structure of DNA and had therefore "discovered the secret of life." While this was a monumental intellectual achievement, the available technology at the time severely limited progress in applying this knowledge. In 1984, Kary Mullis, an organic chemist, figured out a process by which very small quantities of DNA could be amplified with high fidelity. This process, known as polymerase chain reaction (PCR), for the first time, allowed scientists to produce DNA in large quantities. Roughly during this period, Leroy Hood and colleagues at the California Institute of Technology developed four important technologies, DNA and protein sequencers and synthesizers, that automated these processes and led to sequencing of the entire human genome and a discipline now called genomics.

It was at this time that information technology (IT), with its ability to archive, selectively access, and analyze enormous quantities of data, converged with genomics, and its ability to generate the enormous amount of information represented by DNA. The combination of technologies produced an exponential leap in our understanding of genes and what they do that could not have been achieved with the one-gene-at-a-time approach that preceded this convergence. Genomics has led to other "omics," such as proteomics, transcriptomics, and metabolomics (collectively known as panomics), and the application of more sophisticated IT techniques has created the field of systems biology, which, for the first time, considers biological organisms as integrated systems or ecologies. These developments will revolutionize the field of medicine, as the practice of personal medicine—health technologies tailored to the individual—is now possible. Thus, the convergence of organic chemistry, panomics, IT, and medical technology has fundamentally changed modern medical practice, which previously was based on group norms rather than individual differences at the molecular level.

Retrospective studies of convergence are relatively easy to do, and the conclusions from them are fairly clear. It is a different story when looking into the future.

Forecasting

It would be desirable in forecasting to avoid the stovepipe effect. As we have seen, some remarkable advances in technical knowledge and capability have come from the conjunction of developments in two or more disciplines. So, rather than simply projecting each individual discipline, it should be more useful to forecast convergences of disparate areas of S&T. An example is given in figure 1.

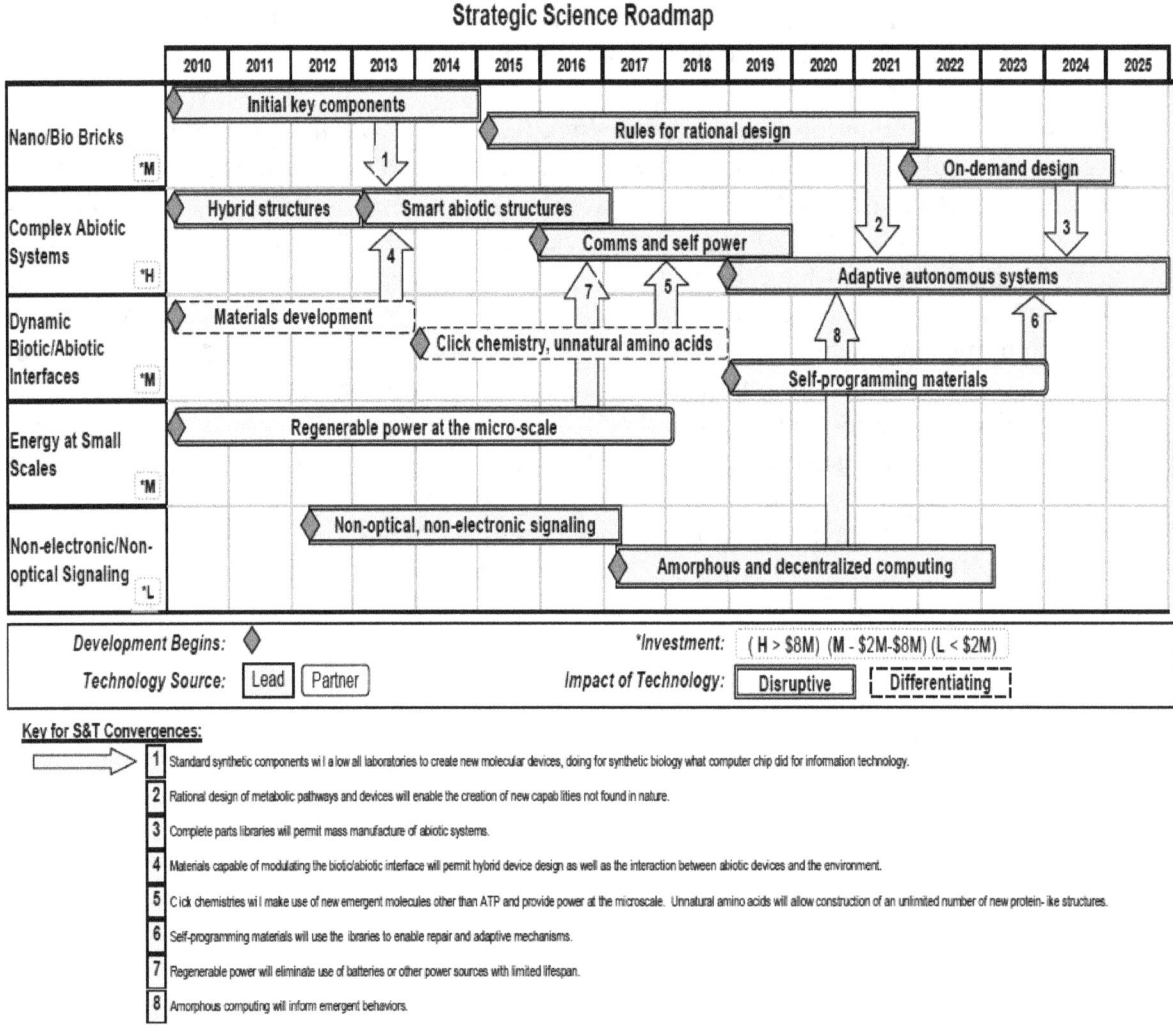

Figure 1. A Roadmap Showing Convergences.

Figure 1 illustrates an example of the projected convergence of technologies which could result in adaptive, autonomous systems possessing some of the characteristics of living organisms. In the left column are four major topic areas. In the rows are critical stages of development within each area, beginning with the year 2010. The vertical arrows connecting various stages of two or more broad areas indicate convergence. For example, in the year 2013, developments in nano/bio bricks (biological building blocks at the atomic or molecular level) join with advances in materials research on biotic/abiotic interfaces. The top of the chart indicates a prediction of convergences from arrows 1 and 4 to enable the creation in abiotic systems of new molecular devices producing hybrids between biological and nonbiological materials.

Farther along on the chart, developments indicated by vertical arrows 2, 6, and 8 are predicted to enable constructing adaptive autonomous systems. The footnotes give the predictions of new capabilities created by combining developments in the separate scientific areas.

Another example is the field of metamaterials. Metamaterials are synthetic composites consisting of a well-ordered, dispersed phase in a matrix; the phases are made of materials of differing properties, such as electromagnetic behavior.[10] First discovered in the 1940s, they became of great interest in the discovery of negative refractive indices for a composite structure comprising a dispersed phase of nanometer dimensions in a matrix. To affect the refractive index, "its features [of the dispersed phase] must be much smaller than the electromagnetic waves utilized. For visible light the composite's structures are of the order of 280nm or less."[11] Such structures are now readily available through the techniques of nanotechnology. Applications may be suggested by the observation in the laboratory of effects known as super lensing and cloaking phenomena. The super lenses are reported to have resolution of as much as three times the diffraction limit. The cloaking devices have the potential to render objects made of certain metamaterials invisible by bending the incoming light wave around the object and then reconstituting the wave on the other side. DARPA has had a program investigating such materials since 2001. While metamaterials and nanotechnology have come together, we should ask what other disciplines will add to the mix to enable new knowledge and new capabilities. Such questions should be addressed in any new forecasting effort.

Finally, the Global Positioning System (GPS) is the result of discoveries in atomic spectroscopy leading to atomic clocks. This precision alone would not have been sufficient to enable creation of the GPS. Also required was the ability to place satellites in desired orbits, which in turn required advances in microelectronics.

Forecasts need to look at discoveries at the frontiers that may give rise to unanticipated new technologies. A review of recent Nobel prizes for science and technology would provide one list of topics.

---

[10] For a current example of the making of such a composite see K.J. Stebe, E. Lewandowski, M. Ghosh, "Oriented Assembly of Metamaterials," *Science*, July 10, 2009, 159–160.

[11] See Wikipedia entry for "Metamaterials," available at <http://en.wikipedia/wiki/Metamaterial>.

Options for New Studies

There are a number of ways to approach organizing and managing forecasts for DOD science and technology. The approaches used in the 1990s studies represent two options. The Army and Navy studies were done on contract to the NRC, while the Air Force study was done by its Scientific Advisory Board (SAB), two very different processes. Studies at the NRC are conducted at arm's length; the process is managed by the NRC, and the services may only offer suggestions. Conversely, the SAB is established by the Air Force and reports to the Chief of Staff and the Secretary, and is thus very close to its sponsor.

Consider now two approaches to a study that could be used. One begins with the basic sciences and related technologies, projects them into the future, and then analyzes the resulting trend lines to see what synergies might occur and what new capabilities might be possible. This has been the traditional approach in forecasting. The other approach is to list future needs for capabilities, and infer from each what technologies and underlying basic scientific advances would be required. In fact, a combination of the two approaches would be the ideal strategy, with the result being a set of charts showing the pathways from the present state of the art out to the new capabilities of the future. The displays may take various forms, two examples being an arrow diagram and a roadmap.

An arrow diagram is easy to develop if one is looking back in time. An example is shown in figure 2, wherein the threads of basic and applied research, advanced development, and engineering development come together to supply the 120mm kinetic energy round known as the A829A1. This is a convergence chart without details for the individual threads that contribute to it. Another way to display the results is to make a roadmap (for an example see figure 1) charting the routes taken by each contributing science or technology as a function of time. This map can be built forward from the present or backward from the future. The map will show when each piece will be ready and which is the rate-limiting step before convergence can occur. A set of research projects can then be designed to mature each thread and funded so as to address the most difficult challenges. The idea of convergences can not only improve forecasting but also be a guide for investing limited S&T funds in key points of leverage.

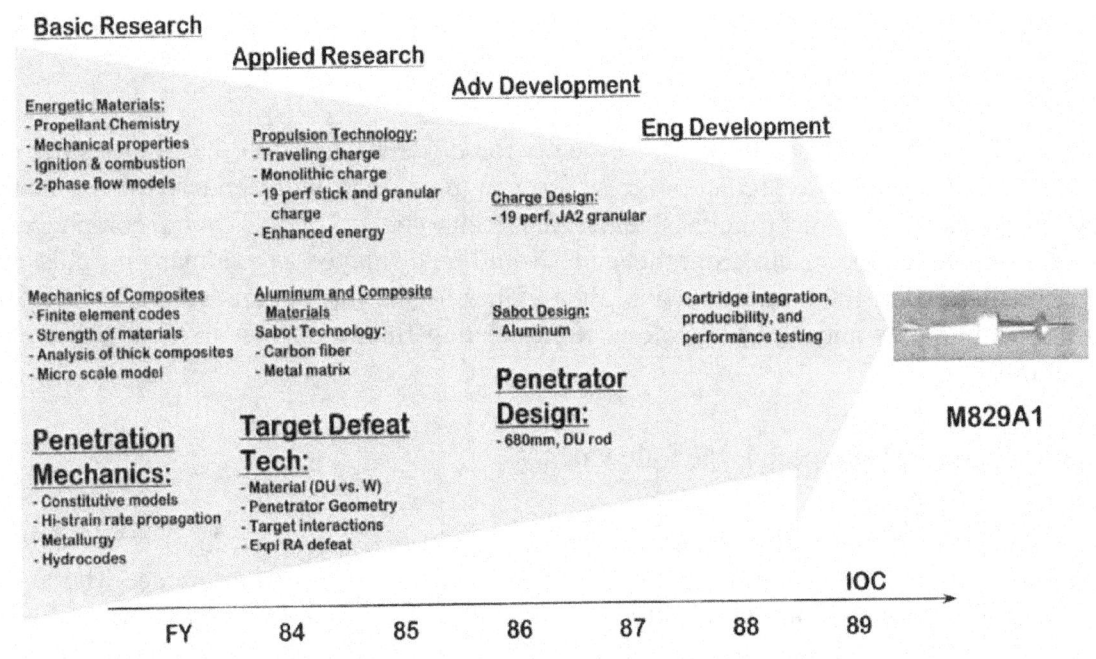

Figure 2. From Basic Research to Final Product – the M829A1 Kinetic Energy Round for the 120mm Gun.[12]

The Need for a Tri-Service Study

The Army's STAR 21 study at the NRC was chartered to identify advances in S&T projected a quarter of a century and more into the future, advances that would be important for ground warfare, and to suggest strategies for developing their full potential.[13] CTNSP, some 16 years later, did an assessment of the predictions made in that study.[14] CTNSP recommended that the Army continue to conduct periodic reviews of critical areas in S&T, making predictions well into the future. The results would help shape the S&T portfolio in those areas as well as give basic information to the senior leadership of the Army and stakeholders elsewhere in the government.

The CTNSP report recommended that such assessments be made every 10 years, that the topics be grouped together, and that the groups be studied on a rolling schedule so that

---

[12] The figure was drawn by the office of the Deputy Assistant Secretary of the Army for Research and Technology for use in various presentations.

[13] See ref. 14. *STAR 21-Strategic Technologies for the Army of the Twenty-First Century* (Washington, D.C.: National Academy Press, 1992).

[14] See ref. 15. John Lyons, Richard Chait, and Jordan Willcox, *An Assessment of the Science and Technology Predictions in the Army's STAR 21 Report*, Defense & Technology Paper 50 (Washington, D.C.: Center for Technology and National Security Policy, July 2008).

only a portion is studied each year. The report recommended that the three armed services carry out the studies jointly, as this would not only provide efficiencies but also draw the services' S&T efforts closer together and help support the needs of the joint operations commands. The report made a number of other suggestions to improve the study process.

Given that the three services will have needs for different end-state capabilities, a tri-service study will need to predict progress for an agreed upon set of underlying sciences, and to predict along the timelines, emerging technologies, devices, and components where applicable. The results from the study would be displayed as roadmaps for clusters of related sciences. Each service would then gather up the threads and convergences for their own purposes and plan technology R&D leading finally to their individual desired capabilities.

The recommended approach is the following:

Each service separately sets forth the capabilities it would like to have 25 (or 15, or 20) years in the future. For example, the Army might have a target for a battlespace where the potential for serious injury to soldiers is nil. Advances in robotics and robotic vehicles, advances in soldier protection, in remote detection and remote fires—all would be applicable.

The three services agree on a list of critical underlying sciences for the study and group them into clusters based on the list of capabilities from the first stage. The clusters may be of closely related sciences, or they may be a mixture of rather different disciplines that might be synergistic. An example of the former could be chemistry, nanoscience, surface science, and material science. An example of the latter is the aforementioned study of nanoscience, biotechnology, information technology, and cognitive science.

Each individual science in a cluster is first projected into the future. Each science would be laid out on a timeline with significant likely advances noted at the projected times. The projections for each member of the cluster would then be plotted at the same scale.

The plots for each cluster are assessed to see where convergences may occur. The convergences are assessed for their potential for something new and unexpected. At some point it becomes clear that the sciences are morphing into technologies with the possibility of conceiving prototype devices and components. At this point the interests of the three services will likely begin to diverge.

Each service plans and budgets in its S&T program accordingly. Each service would then assess the potential of following up the most promising leads from the predictions for the sciences and the clusters of sciences.

Organizing and Managing the Studies

There are several different ways to conduct a tri-service study of this kind. The questions to be answered include: What is the scope of the study? Who should sponsor the study? Who should set the agenda? Who should manage the study? Who should serve on the expert panel(s), and how should they be selected? What should be the role of DOD in-house experts? Each of these questions is addressed below.

The first decision is whether to restrict the tri-service study to the prediction of developments in basic science. Going beyond basic science to more applied work to explore new devices and components could be done jointly if the work stops short of considering complete systems, since such systems will differ from service to service. The study should be funded jointly by the three services or by the DDR&E office working with the services. DDR&E participation would add impact on senior-level decisionmaking once the results are available. The agenda for the study would consist of an assessment and projection into the future of a set of scientific disciplines. These would be done in clusters of disciplines thought likely to produce synergies. The agenda would be set by senior DOD scientists selected by the Defense S&T Advisory Group (DSTAG) working with staff from the DDR&E office and the services.

Next is the decision as to who should manage the study. One option is to do the study in-house using in-house experts. Another possibility is to contract the project to an outside group. Two of the three studies done in the 1990s were conducted by the NRC; the third was done by the Air Force SAB, a formal advisory committee under the Federal Advisory Committee Act. Other options include such analytical organizations as RAND and MITRE. Contracting for the study would give the results more credibility than if it is done solely in-house, the latter tending to be criticized as being too limited and too parochial. Maximum credibility would be obtained by using the NRC.

The study should be made by a set of expert panels, perhaps one for each cluster of sciences. It would be possible to have a panel for each science, but this would be unwieldy and would increase the chance of missing possible convergences. Using a panel per cluster is a more economical approach and would get at the objective of finding convergences more effectively. For assessing clusters of sciences, the panels should have experts from each science in the cluster. Additionally, one or more generalists and futurists would be useful. These panelists should not be too focused on their own subdisciplines but rather should have broad knowledge of their parent disciplines. Experience shows that if a panel member has too strong a bias toward his or her own work, the work of the panel can be skewed. Pains should be taken to avoid or manage conflicts of interest. This should be done by requiring a statement of potential issues by the nominees, thereby enabling the contractor either to disqualify the nominee or to make appointments to balance the conflicts. In the latter procedure, the nominees should publicly identify their biases. This disclosure should also indentify sources of sponsored funds to each panelist for his or her own research.

13

Finally, the study should make use of in-house DOD subject matter experts (SMEs). In past NRC studies, SMEs from the government sponsor have been relegated to observer status with the possibility of answering questions only to clarify some aspect of the sponsor's work. A more active role of DOD SMEs would give the panels a fuller appreciation of the military's interest and needs. Certainly the military's most senior technical staff—especially the people holding ST rank (a non-administrative rank for Senior Technical professionals equivalent to members of the Senior Executive Service and flag officers)—would add strength and perspective to panel deliberations. Their participation would also help them to educate the DOD staff who must deal with the results of the study, and would provide corporate memory as the recommendations are implemented.

Conclusion

Forecasting science and technology is a worthwhile exercise in that it sharpens planning in the shorter term for, among other things, aligning the research program with the TRADOC list of needed capabilities, selectively strengthen some areas while reducing others, and adjusting the research and development budget. We believe forecasting should be done by panels of leading experts in the fields to be studied and that selected disciplines should be studied together to look for potential convergences or serendipities. We also believe and strongly recommend that DOD conduct tri-service studies of underlying sciences important to the military, and that such studies be done in a rolling fashion so that the scope not be overly ambitious for any one year. We conclude that any given area should be the subject of a forecast every 10 years. Such studies will result in more convincing program rationales and should engender more support from senior DOD leaders and the Congress.